BEYOND THE VEIL: EXPLORING THE BUBBLE THEORY AND THE FOURTH DIMENSION

A Beginner's Journey into Higher Dimensions and Modern Physics

Jameel Chamberlain

Copyright © 2024 Jameel Chamberlain

All rights reserved

No part of this book may be reproduced, stored in a retrieval system, or transmitted in any form or by any means, electronic, mechanical, photocopying, recording, or otherwise, without express written permission of the author.

Cover design by: Jameel Chamberlain

Printed in the United States of America

To all the curious minds who dare to explore the unknown.

CONTENTS

Title Page
Copyright
Dedication
Introduction — 2
Chapter 1: Life in Three Dimensions — 8
Chapter 2: The Fourth Dimension—A New Frontier — 14
Chapter 3: The Building Blocks of the Universe — 20
Chapter 4: Beyond the Visible—Dark Matter and Energy — 26
Chapter 5: The Bubble Theory Simplified — 32
Chapter 6: The 'S' Energy Field Explained — 38
Chapter 7: Bubbles in the Fourth Dimension — 44
Chapter 8: When Bubbles Interact — 50
Chapter 9: Rethinking Gravity and the Forces — 56
Chapter 10: Shedding Light on Dark Matter and Energy — 62
Chapter 11: Quantum Connections — 70
Chapter 12: Philosophical Reflections — 78
Chapter 13: The Journey of Discovery — 86
Chapter 14: The Future Horizons — 92
Glossary — 98
Further Reading and Resources — 100
About The Author — 102

Acknowledgement 104

INTRODUCTION

A Sky Full of Questions

On a clear night, away from the city lights, the sky unfolds like a vast canvas sprinkled with countless stars. Imagine standing in an open field, the cool breeze gently rustling the grass around you. Above, the moon hangs like a silver lantern, its soft glow casting a serene light on the landscape. For millennia, humans have gazed up at this breathtaking sight, filled with wonder and curiosity. The stars, twinkling like distant beacons, have always held a special place in our hearts and minds.

Our ancestors looked to the stars not just for their beauty but for guidance. The night sky was their map and calendar, helping them navigate vast oceans and predict the changing seasons. Constellations told stories of heroes, gods, and mythical creatures, weaving a tapestry of lore that connected cultures across the globe. The moon, with its ever-changing phases, marked the passage of time, symbolizing renewal and the cyclical nature of life.

But beyond practical uses, the sky has always stirred deeper questions within us. What are these distant lights? How far away are they? Are we alone in this immense expanse? Standing under that canopy of stars, one can't help but feel both insignificant and profoundly connected to something greater.

A Journey of Curiosity

From the earliest cave paintings to modern telescopes peering into the depths of space, our desire to understand the cosmos has been a driving force in human history. Ancient astronomers in civilizations like Babylon, Egypt, and China meticulously recorded celestial events, seeking patterns and meanings in the movements of the heavens.

The Greek philosopher Aristotle envisioned a geocentric universe, with Earth at its center, surrounded by concentric celestial spheres. This view held sway for centuries until the revolutionary ideas of Nicolaus Copernicus in the 16th century proposed a heliocentric model, placing the Sun at the center. This bold shift challenged the established order and set the stage for a new era of exploration.

Galileo Galilei, with his improved telescope, observed moons orbiting Jupiter and the phases of Venus, providing tangible evidence that not everything revolved around the Earth. His discoveries ignited a scientific revolution, paving the way for Isaac Newton's laws of motion and universal gravitation, which mathematically described the forces governing celestial bodies.

Fast forward to the 20th century, Albert Einstein's theories of relativity reshaped our understanding of space and time, revealing a universe far more complex and fascinating than previously imagined. The cosmos was no longer a static backdrop but a dynamic entity, expanding, curving, and interacting in ways that continue to intrigue scientists today.

Unanswered Questions and New Frontiers

Despite these incredible advancements, many mysteries remain. What is the true nature of gravity? How do we reconcile the seemingly incompatible theories of general relativity and quantum mechanics? What is the composition of the universe, most of which appears to be made of invisible dark matter and dark energy?

As we delve deeper into these questions, new theories emerge, challenging our perceptions and pushing the boundaries of knowledge. Concepts like extra dimensions, parallel universes, and the multiverse enter the conversation, offering tantalizing possibilities.

Guided by the Stars, Inspired by the Moon

Just as our ancestors looked to the stars for direction, we too seek guidance in the vast unknown. The moon, transitioning through its phases, reminds us of the ever-changing nature of understanding. Each new discovery is a step in an ongoing journey—a transition from old ideas to new horizons.

Consider the moon landing in 1969. When Neil Armstrong set foot on the lunar surface, it wasn't just a monumental technological achievement; it was a profound moment that united humanity in wonder. The famous words, "That's one small step for man, one giant leap for mankind," encapsulated the spirit of exploration and the limitless potential of human ingenuity.

The Eternal Quest

Our fascination with the cosmos is more than a pursuit of knowledge; it's a fundamental part of who we are. The night sky connects us to our past, ignites our imagination, and beckons us to explore. Each star is a reminder of the countless possibilities that await discovery.

In this book, we embark on a journey to explore one such possibility—the Bubble Theory. It's an idea that delves into the very fabric of reality, suggesting that there may be more dimensions and layers to the universe than we can perceive.

We'll venture beyond the familiar three dimensions, exploring concepts that challenge conventional wisdom. Through simple analogies, thought experiments, and accessible explanations, we'll unlock the doors to higher dimensions and consider how they might explain some of the universe's greatest mysteries.

An Invitation to Wonder

As you turn the pages, I invite you to rekindle that childlike curiosity—the awe felt when first gazing at the stars. Let's ponder big questions together, embrace the unknown, and perhaps, in the process, glimpse a new understanding of the universe we call home.

Whether you're a seasoned science enthusiast or simply someone intrigued by the wonders of existence, this journey is for you. Let's look up once more at the night sky, allowing our imaginations to soar among the stars and beyond, into the uncharted territories of the fourth dimension.

Key Takeaways from Introduction:

- **Human Curiosity:** From ancient times to the present, humans have looked to the stars and the moon for guidance, inspiration, and understanding.
- **Progress of Astronomy:** Significant figures like Copernicus, Galileo, Newton, and Einstein have revolutionized our understanding of the cosmos.
- **Ongoing Mysteries:** Despite advancements, many questions about the universe remain unanswered, prompting new theories and explorations.
- **The Bubble Theory:** Introduced as a new perspective to explore, potentially offering insights into higher dimensions and the fabric of reality.
- **Invitation to Explore:** Readers are encouraged to embark on this journey with an open mind and a sense of wonder.

CHAPTER 1: LIFE IN THREE DIMENSIONS

Our Everyday Reality

Take a moment to look around you. Whether you're sitting in a cozy living room, walking down a bustling street, or enjoying a serene park, your experience is shaped by three dimensions: length, width, and height. These dimensions define the space we inhabit and navigate every day. They're so fundamental to our existence that we rarely stop to consider them—they simply are.

When you reach out to grab a cup of coffee, you're moving your hand through these three dimensions. Architects design buildings considering all three to create functional and aesthetically pleasing spaces. Athletes train to excel in movement within this tri-dimensional arena. Our entire physical world is built upon the fabric of these three dimensions.

Defining Dimensions in Simple Terms

But what exactly is a dimension? In the simplest terms, a dimension is a direction in which you can move or measure something. Let's break it down:

- **Length (X-axis):** This is the dimension that runs left to right. Imagine standing on a straight road stretching out to the horizon—that's length.
- **Width (Y-axis):** This dimension runs front to back. Think of a path that crosses your road, allowing you to move forward or backward.
- **Height (Z-axis):** The dimension that goes up and down.

Consider climbing a flight of stairs or a bird soaring into the sky.

These three dimensions are perpendicular to each other, meaning they all intersect at right angles. Together, they form the coordinate system that maps out our physical space.

Flatland: A World of Two Dimensions

To understand dimensions better, let's embark on a thought experiment inspired by Edwin A. Abbott's classic novella, *Flatland*. Imagine a world that exists in only two dimensions—a flat plane with only length and width, like a sheet of paper. The inhabitants of this world, let's call them Flatlanders, are shapes like circles, squares, and triangles. They can move left and right, forward and backward, but the concept of "up" and "down" is entirely foreign to them.

For a Flatlander, everything is experienced in two dimensions. If you were to place a finger on their world, they would perceive it as a sudden, inexplicable shape appearing out of nowhere. They have no awareness of the third dimension, just as we may lack awareness of dimensions beyond our own.

Imagine a New Direction

Now, consider how we, as three-dimensional beings, can comprehend the idea of a fourth spatial dimension. It's challenging because all our experiences are confined to length, width, and height. But just as a Flatlander might struggle to imagine "up" and "down," we can attempt to stretch our minds to envision a direction perpendicular to all three of our known dimensions.

Think of it like this: If you were to draw a dot on a piece of paper and then draw a line extending from it, you've added a dimension—moving from zero dimensions (the dot) to one dimension (the line). Add another line perpendicular to the first, and you've created a two-dimensional plane. Introduce a line perpendicular to both of those, and you have a three-dimensional space. Now, imagine adding yet another line perpendicular to all three existing ones. This is where the concept becomes abstract because we can't physically visualize it, but mathematically, it can be described.

Perception Limits

Our senses are tailored to perceive the three dimensions necessary for survival. Our eyes capture light that reflects off objects in our environment, allowing us to judge distance, depth, and movement within three-dimensional space. However, we can't directly perceive anything beyond these dimensions. Just as a colorblind person might not experience certain hues, or a person can't hear frequencies beyond their auditory range, we may be blind to aspects of reality existing in higher dimensions.

Shadows and Projections

One way to approach understanding higher dimensions is by examining how higher-dimensional objects cast shadows in lower dimensions. For instance, a three-dimensional object, like a ball, casts a two-dimensional shadow on a wall. The shadow changes shape depending on the angle of the light, but it can never fully represent the ball's three-dimensional form.

Similarly, a four-dimensional object would cast a three-dimensional "shadow" in our world. This projection might seem bizarre or impossible to interpret because we're missing the full picture.

The Tesseract: A Glimpse into the Fourth Dimension

A common example used to visualize the fourth dimension is the tesseract, or hypercube. Just as a cube is a three-dimensional extension of a square, a tesseract is a four-dimensional extension of a cube.

- **Point (0D):** No dimensions.
- **Line (1D):** Two points connected, extending in one dimension.
- **Square (2D):** Four lines connected, extending in two dimensions.
- **Cube (3D):** Six squares connected, extending in three dimensions.
- **Tesseract (4D):** Eight cubes connected, extending in four dimensions.

When we attempt to represent a tesseract in three dimensions, we create a projection that appears as a cube within a cube, with lines connecting the corresponding vertices. It looks distorted because we're compressing a higher-dimensional object into lower-dimensional space, losing some of its properties in the process.

Why Consider Higher Dimensions?

You might wonder why we should bother thinking about dimensions beyond the three we experience. The answer lies in the pursuit of understanding the fundamental nature of the universe. In physics, higher dimensions are not just abstract concepts but potential explanations for complex phenomena.

- **String Theory:** Proposes that additional dimensions exist but are compactified or curled up at scales too small for us to detect. These extra dimensions could unify the forces of nature or explain the behavior of fundamental particles.
- **Unified Theories:** Higher dimensions might provide the framework necessary to reconcile general relativity with quantum mechanics.

An Invitation to Imagine

While we may not be able to perceive higher dimensions directly, we can use mathematics, analogies, and imagination to explore their possibilities. Just as Flatlanders might theorize about a third dimension, we can speculate about the fourth and beyond.

As we journey further into this book, we'll delve into how these higher dimensions might interact with our own, potentially revealing new insights into the mysteries of the universe. We'll explore the Bubble Theory, which suggests that our reality might be just one of many coexisting within a higher-dimensional space.

So, let's keep our minds open and our imaginations active as we step into realms beyond the familiar. After all, every great discovery begins with the willingness to see beyond the obvious.

Key Takeaways from Chapter 1:

- **Dimensions Defined:** Dimensions are directions in which movement or measurement is possible—length, width, and height define our three-dimensional space.
- **Flatland Analogy:** Imagining a two-dimensional world helps us conceptualize how higher dimensions might exist beyond our perception.
- **Visualizing Higher Dimensions:** Tools like projections and analogies allow us to consider four-dimensional objects, even if we can't perceive them directly.
- **Relevance to Physics:** Higher dimensions are integral to theories attempting to explain fundamental aspects of the universe, such as string theory.
- **Embracing Imagination:** Considering higher dimensions requires us to think creatively and step beyond the limits of our senses.

CHAPTER 2: THE FOURTH DIMENSION— A NEW FRONTIER

Time as the Fourth Dimension

When we hear "the fourth dimension," many of us think of time. In Einstein's theory of relativity, time is intertwined with the three spatial dimensions, forming a four-dimensional fabric called spacetime. This concept revolutionized physics, showing that time can dilate and space can contract depending on an object's relative motion.

But what if there's more to the fourth dimension than time? What if there's a fourth spatial dimension—a direction perpendicular to all three of our known spatial dimensions? This idea stretches our imagination and challenges our understanding of reality.

Beyond Spacetime

In the realm of theoretical physics, the possibility of extra spatial dimensions is not just a flight of fancy but a serious consideration. Theories like string theory and M-theory propose the existence of additional dimensions to unify the fundamental forces of nature.

- **String Theory:** Suggests that the fundamental particles we observe are not point-like dots but tiny vibrating strings. These strings' vibrations correspond to different particles, and they require additional spatial dimensions to be mathematically consistent—up to ten or eleven dimensions in some models.
- **Compactification:** The idea that extra dimensions might

be "curled up" at incredibly small scales, making them imperceptible at human scales.

Visualizing the Fourth Spatial Dimension

Visualizing a fourth spatial dimension is no easy task, but let's attempt it using analogies and projections.

Analogy of the Shadow

Consider a three-dimensional object casting a two-dimensional shadow. A cube, for example, can cast a square shadow, but the shadow doesn't capture all the cube's properties. Similarly, a four-dimensional object would cast a three-dimensional "shadow" in our world.

The Hypersphere

Imagine a sphere expanding in three dimensions. Now, think of a hypersphere—a four-dimensional sphere—expanding in the fourth spatial dimension. From our perspective, we might observe a three-dimensional sphere appearing out of nowhere, growing, then shrinking, and disappearing, much like how a sphere passing through Flatland would appear to the Flatlanders.

The Hypercube (Tesseract) Revisited

As introduced in the previous chapter, the tesseract is a four-dimensional analogue of the cube. When we try to represent it in three dimensions, we see a cube within a cube, connected at the corners. This projection helps us grasp the concept, even if imperfectly.

Implications of a Fourth Spatial Dimension

If a fourth spatial dimension exists, it could have profound implications:

- **New Physics:** Forces and particles might behave differently when higher dimensions are considered, potentially

resolving inconsistencies in current theories.
- **Gravity's Weakness:** Gravity is significantly weaker than the other fundamental forces. Some theories suggest that gravity may propagate through extra dimensions, diluting its strength in our three-dimensional space.
- **Parallel Universes:** The idea that our universe is one of many existing in higher-dimensional space, each possibly with its own laws of physics.

Everyday Analogies

To bring this abstract concept closer to home, let's consider some analogies:

The Book Pages

Think of our universe as a single page in a vast book. The other pages represent parallel universes existing alongside ours, separated by the "thickness" of the fourth dimension. While the characters on one page are unaware of those on another, all pages coexist within the same book.

The Ant on a Tightrope

Imagine an ant walking along a tightrope. To us, the tightrope is a one-dimensional line. However, the ant can move not only forward and backward but also around the circumference of the rope—a second dimension hidden from our larger perspective. Similarly, extra dimensions might be compact and unnoticed at our scale but significant at smaller scales.

Experimental Searches for Extra Dimensions

Scientists have devised experiments to detect the presence of extra dimensions:

- **Particle Accelerators:** High-energy collisions, like those in the Large Hadron Collider (LHC), could produce particles that momentarily escape into extra dimensions, resulting in

missing energy signatures.
- **Precision Measurements of Gravity:** At very short distances, gravity might deviate from Newton's inverse-square law if extra dimensions are influencing it. Experiments attempt to measure gravitational attraction at micrometer scales.

Challenges and Skepticism

While the mathematics of theories involving extra dimensions can be compelling, they face challenges:

- **Lack of Direct Evidence:** Despite extensive efforts, no experimental data has conclusively proven the existence of extra spatial dimensions.
- **Complexity:** The theories can be mathematically intricate, making them difficult to test and verify.
- **Alternative Explanations:** Some phenomena attributed to extra dimensions might be explained by other theories, such as modifications to gravity or undiscovered particles.

Keeping an Open Mind

Science progresses by exploring bold ideas and rigorously testing them. The concept of a fourth spatial dimension pushes the boundaries of our understanding and invites us to think creatively.

As we continue our journey, we'll delve into the Bubble Theory—a new perspective on how a fourth spatial dimension might manifest and influence our reality. We'll explore how this idea could provide insights into some of the universe's most perplexing mysteries, all while keeping our explanations accessible and grounded in relatable concepts.

Key Takeaways from Chapter 2:

- **Extra Spatial Dimensions:** The possibility of dimensions beyond the three we experience is a serious consideration in

theoretical physics.
- **Visualization Challenges:** Using analogies like shadows, hyperspheres, and projections helps us conceptualize higher dimensions.
- **Implications for Physics:** A fourth spatial dimension could explain phenomena like gravity's relative weakness and the behavior of fundamental particles.
- **Experimental Efforts:** Scientists are actively searching for evidence of extra dimensions through particle physics and gravitational experiments.
- **Open-Minded Exploration:** Embracing new ideas, even those that challenge our perceptions, is essential for advancing scientific understanding.

CHAPTER 3: THE BUILDING BLOCKS OF THE UNIVERSE

Atoms: The Invisible Architecture

At the heart of everything we see, touch, and experience are atoms—the fundamental units of matter. Each atom consists of a nucleus made of protons and neutrons, surrounded by a cloud of electrons. These tiny structures come together in countless combinations to form the molecules that make up everything from the air we breathe to the stars in the sky.

Consider a single drop of water. It contains approximately 1.5×10^{21} water molecules, each made of two hydrogen atoms bonded to one oxygen atom. Despite their minuscule size, atoms are incredibly significant, forming the diverse tapestry of the material world.

Particles Beyond the Atom

Delving deeper, we find that protons and neutrons are themselves composed of even smaller particles called quarks, held together by gluons. Electrons, quarks, and gluons are part of a family known as elementary particles, which, as far as we know, are not made of anything smaller.

These particles interact through fundamental forces, governing the behavior of matter at the smallest scales. Understanding these interactions is key to unlocking the secrets of the universe.

The Fundamental Forces Simplified

There are four known fundamental forces in nature:

Gravity

- **What it Does:** Attracts objects with mass toward one another.
- **Everyday Experience:** Keeps us grounded on Earth, governs the motion of planets and stars.
- **Unique Feature:** Although it has an infinite range, gravity is the weakest of the forces at the particle level.

Electromagnetism

- **What it Does:** Acts between electrically charged particles.
- **Everyday Experience:** Powers electrical devices, enables magnets to stick to refrigerators.
- **Unique Feature:** Responsible for light and all electromagnetic radiation.

Strong Nuclear Force

- **What it Does:** Holds protons and neutrons together in the nucleus.
- **Everyday Experience:** Not directly felt but crucial for the stability of matter.
- **Unique Feature:** The strongest force but operates only at very short ranges within the nucleus.

Weak Nuclear Force

- **What it Does:** Responsible for radioactive decay and certain nuclear reactions.
- **Everyday Experience:** Plays a role in processes that power the sun.
- **Unique Feature:** Affects subatomic particles and also has a short range.

Fields: The Invisible Influence

A field is a region in space where a force has an effect. Think of it as an invisible landscape that tells particles how to move.

- **Gravitational Field:** Determines how objects with mass attract each other.
- **Electromagnetic Field:** Governs how charged particles interact.
- **Quantum Fields:** In quantum physics, every particle is associated with a field that permeates space.

An easy way to visualize a field is to sprinkle iron filings around a magnet. The filings align along the magnetic field lines, revealing the otherwise invisible influence.

Waves and Duality

Particles can exhibit both particle-like and wave-like properties, a concept known as wave-particle duality.

- **Light as a Wave:** Exhibits interference and diffraction patterns.
- **Light as a Particle:** Photons deliver energy in discrete packets.
- **Electrons and Other Particles:** Also display wave-like behavior, as demonstrated in the famous double-slit experiment.

The Uncertainty Principle

At quantum scales, the exact position and momentum of a particle cannot be simultaneously known. This uncertainty isn't due to measurement flaws but is inherent to nature.

- **Implications:** Leads to phenomena like quantum tunneling, where particles can pass through barriers they seemingly shouldn't be able to.
- **Everyday Analogy:** Imagine trying to measure the exact speed and position of a hummingbird's wings mid-flap—it's not just difficult; it's fundamentally uncertain at that scale.

Limitations of Current Models

While quantum mechanics and general relativity are highly successful in their respective domains, they are fundamentally incompatible when it comes to describing certain phenomena, such as what happens inside black holes or the conditions at the universe's inception.

- **The Need for Unification:** Physicists seek a theory of quantum gravity that would reconcile these two pillars of physics.
- **String Theory and Beyond:** Proposals involving extra dimensions aim to bridge this gap.

Opening the Door to New Ideas

Understanding the building blocks of the universe sets the stage for exploring theories that push the boundaries of conventional physics. The Bubble Theory is one such idea, proposing that additional dimensions and structures could provide answers to unresolved questions.

As we continue our journey, we'll build upon this foundation to explore how the Bubble Theory might integrate with or extend these fundamental concepts, offering new perspectives on the nature of reality.

Key Takeaways from Chapter 3:

- **Atoms and Particles:** Matter is composed of atoms, which consist of smaller particles like protons, neutrons, and electrons.
- **Fundamental Forces:** Four forces govern interactions in the universe—gravity, electromagnetism, strong nuclear force, and weak nuclear force.
- **Fields and Waves:** Fields influence how particles move, and particles exhibit both wave-like and particle-like properties.
- **Quantum Mechanics:** Introduces principles like wave-particle duality and uncertainty, revealing the complex

nature of the microscopic world.
- **Challenges in Physics:** Current models have limitations, highlighting the need for new theories that can unify our understanding.

CHAPTER 4: BEYOND THE VISIBLE—DARK MATTER AND ENERGY

The Cosmic Puzzle

When we observe the universe, from the rotation of galaxies to the expansion of space itself, we encounter phenomena that can't be fully explained by the matter we can see. Astronomers have discovered that the visible matter—stars, planets, gas, and dust—accounts for only a small fraction of the universe's total mass and energy.

Dark Matter: The Invisible Scaffold

Evidence for Dark Matter

- **Galaxy Rotation Curves:**
    - **Observation:** Stars at the edges of galaxies rotate at similar speeds to those near the center, defying expectations based on visible mass.
    - **Implication:** There must be additional unseen mass exerting gravitational influence.
- **Gravitational Lensing:**
    - **Observation:** Light from distant objects bends around massive clusters in ways that suggest more mass than is visible.
    - **Implication:** Invisible matter is affecting the path of light.
- **Cosmic Microwave Background (CMB):**
    - **Observation:** Patterns in the CMB radiation provide

a snapshot of the early universe, indicating more mass than can be accounted for by visible matter.

What Is Dark Matter?

- **Unknown Composition:** Dark matter doesn't emit, absorb, or reflect light, making it undetectable by conventional means.
- **Possible Candidates:**
  - **WIMPs (Weakly Interacting Massive Particles):** Hypothetical particles that interact through gravity and possibly the weak nuclear force.
  - **Axions, Neutralinos, and Other Exotic Particles:** Proposed by various theories but not yet observed.

Analogy: The Wind You Can't See

Imagine standing outside on a windy day. You can't see the wind, but you feel it on your skin, see leaves rustling, and hear it whistle through the trees. Similarly, dark matter is invisible but reveals its presence through gravitational effects on visible matter.

Dark Energy: The Mysterious Force of Expansion

Discovery of Dark Energy

In the late 1990s, observations of distant supernovae revealed that the universe's expansion is accelerating, not slowing down as previously thought.

- **Accelerating Expansion:** Galaxies are moving away from each other at increasing speeds.
- **Implication:** There's a force counteracting gravity on cosmic scales.

What Is Dark Energy?

- **Unknown Nature:** Dark energy is even more mysterious

than dark matter.
- **Possible Explanations:**
  - **Cosmological Constant (Λ):** A term in Einstein's equations representing energy inherent to space itself.
  - **Quintessence:** A dynamic field that changes over time and space.

Analogy: Rising Dough

Imagine baking bread. As the dough rises, raisins embedded within move away from each other. If the dough begins to rise faster over time, the raisins' separation accelerates. Dark energy is like an unknown ingredient causing the dough (space) to expand more rapidly.

The Cosmic Energy Budget

- **Ordinary Matter:** Approximately 5% of the universe.
- **Dark Matter:** About 27%.
- **Dark Energy:** Roughly 68%.

This means that about 95% of the universe is made up of components we cannot directly observe.

Challenges to Understanding

- **Detection Difficulties:** Dark matter and dark energy don't interact with light, making them elusive.
- **Theoretical Gaps:** Existing theories don't fully explain their properties or origins.
- **Competing Models:** Alternative theories, like modified gravity, attempt to account for observations without invoking dark matter or energy.

Why It Matters

Understanding dark matter and dark energy is crucial for a complete picture of the universe's structure, evolution, and ultimate fate. These components influence the formation of galaxies, the behavior of cosmic structures, and the long-term dynamics of space itself.

Opening the Door to the Bubble Theory

The unexplained nature of dark matter and dark energy invites fresh perspectives. The Bubble Theory offers a new way to think about these phenomena:

- **Overlapping Bubbles:** Perhaps the gravitational effects we attribute to dark matter are due to interactions with other "bubbles" in the fourth dimension.
- **'S' Energy Field Influence:** Dark energy might be a manifestation of dynamics within the 'S' Energy Field affecting the expansion of our universe.

As we explore these ideas in the following chapters, we'll consider how the Bubble Theory might provide insights into these cosmic mysteries, potentially offering explanations that bridge gaps in our current understanding.

Key Takeaways from Chapter 4:

- **Dark Matter:** An unseen form of matter inferred from gravitational effects on visible matter and light.
- **Dark Energy:** A mysterious force driving the accelerated expansion of the universe.
- **Cosmic Composition:** Ordinary matter makes up only a small fraction of the universe's total content.
- **Scientific Significance:** Unraveling the nature of dark matter and energy is essential for a comprehensive understanding of cosmology.
- **Invitation to New Ideas:** The mysteries of dark matter and dark energy open the door for theories like the Bubble Theory to offer alternative explanations.

CHAPTER 5: THE BUBBLE THEORY SIMPLIFIED

A New Perspective on Reality

Imagine the universe as an immense ocean, vast and deep. Within this ocean, countless bubbles rise and fall, each one a world unto itself. These bubbles represent independent realms, coexisting within the same space yet remaining distinct due to their unique properties. This is the essence of the Bubble Theory—a new way to understand the fabric of reality and the possible existence of higher dimensions.

In this chapter, we'll explore the Bubble Theory in simple terms, using analogies and examples to make the concept accessible. We'll delve into how these bubbles can exist simultaneously without interfering with one another and what binds them together in the fourth dimension.

The Radio Analogy: Tuning into Different Realities

Consider how a radio works. In any given place, numerous radio signals fill the air, each broadcasting on its own frequency. When you tune your radio to a specific frequency, you access that particular station, while the others remain undetected. All these signals coexist in the same space without interfering because they operate at different frequencies.

Similarly, the Bubble Theory suggests that multiple universes —or "bubbles"—exist within the same spatial framework but are differentiated by unique "frequencies" known as 'S' energy signatures. Just as radio frequencies allow multiple stations to broadcast simultaneously, the 'S' energy signatures enable multiple bubbles to occupy the same space without interacting.

Understanding the 'S' Energy Signature

At the heart of the Bubble Theory is the concept of the 'S' Energy Field and its unique signatures. Think of the 'S' energy signature as a specific vibration or pattern that defines each bubble, much like a fingerprint or a musical note.

- **Unique Identifier:** The 'S' energy signature is what makes each bubble distinct, preventing it from merging with others.
- **Binding Force:** It acts as a glue, holding the matter within a bubble together in the fourth dimension.
- **Non-Interference:** Because each bubble has a unique signature, they don't interfere with one another, allowing for peaceful coexistence.

Bubbles Coexisting in the Fourth Dimension

To visualize how bubbles can occupy the same space without interacting, let's use the analogy of transparent soap bubbles. Imagine blowing several soap bubbles that float and sometimes overlap. Because they're transparent, you can see through them, and when they overlap, each bubble maintains its own surface tension, keeping it intact.

In the Bubble Theory, each bubble represents a universe with its own set of physical laws and matter, bound by its 'S' energy signature. Even if these bubbles overlap in space, their unique signatures prevent them from interacting directly.

Layers of Reality: The Book Pages Analogy

Another way to think about this is to imagine a book. Each page in the book represents a different bubble, a separate reality.

The pages are stacked together, occupying the same overall space (the book), but the characters on one page are unaware of those on another. The 'S' energy signatures are like the page numbers, keeping each story distinct and organized.

Why Don't We Perceive Other Bubbles?

You might wonder why we don't notice these other bubbles if they occupy the same space as ours. The answer lies in our sensory limitations and the uniqueness of the 'S' energy signatures.

- **Sensory Limitations:** Our senses and instruments are tuned to detect phenomena within our own bubble's 'S' energy signature.
- **Different Frequencies:** Just as a radio tuned to one frequency doesn't pick up other stations, we're "tuned" to our bubble's signature.
- **Dimensional Constraints:** Our perception is confined to three spatial dimensions, making it challenging to detect or interact with phenomena in higher dimensions.

An Everyday Example: Multiple Conversations

Imagine being at a crowded party where many conversations happen simultaneously. You can focus on the person you're talking to, while other conversations fade into the background noise. Each conversation is like a bubble with its own 'S' energy signature. While they're all occurring in the same room, they don't interfere with your ability to communicate with your conversation partner.

The Non-Interference Principle

A key aspect of the Bubble Theory is the non-interference principle. Because each bubble has a unique 'S' energy signature, they remain independent even when occupying the same space.

This principle ensures that:

- **Structural Integrity:** Each bubble maintains its own physical laws and structures without being affected by others.
- **Energy Isolation:** Energy and matter within one bubble don't transfer to another under normal circumstances.
- **Stability:** The uniqueness of signatures prevents chaotic interactions that could destabilize the bubbles.

Possibility of Interaction: A Glimpse into the Extraordinary

While the non-interference principle keeps bubbles separate, the Bubble Theory also introduces the intriguing possibility of interaction under special conditions.

- **Interlinking Bubbles:** If two bubbles' 'S' energy signatures become sufficiently similar, they might begin to interact.
- **Energy Fluctuations:** External influences or internal fluctuations could alter a bubble's signature, leading to potential overlap.
- **Phenomenological Implications:** Such interactions could explain certain unexplained phenomena or anomalies in our universe.

We'll explore these ideas further in subsequent chapters, delving into how interlinking might occur and what it could mean for our understanding of reality.

Embracing the Unknown

The Bubble Theory invites us to expand our minds and consider possibilities beyond our everyday experience. By using analogies and familiar concepts, we can begin to grasp how multiple realities might coexist and how the fourth dimension plays a crucial role in this framework.

As we continue our journey, we'll examine the 'S' Energy Field in more detail, explore the nature of bubbles in the fourth dimension, and discuss the implications of this theory for physics and cosmology.

Key Takeaways from Chapter 5:

- **Bubble Theory Overview:** Proposes that multiple universes (bubbles) exist simultaneously in the same space, differentiated by unique 'S' energy signatures.
- **'S' Energy Signature:** Acts as a unique identifier and binding force for each bubble, preventing interference with others.
- **Analogies Used:**
    - **Radio Frequencies:** Different stations (bubbles) coexist without interference due to unique frequencies ('S' energy signatures).
    - **Soap Bubbles:** Overlapping bubbles maintain integrity due to their unique surface tension.
    - **Book Pages:** Separate stories (bubbles) coexist within the same book (space) without characters interacting.
- **Non-Interference Principle:** Ensures that bubbles remain independent even when occupying the same space.
- **Possibility of Interaction:** Under special conditions, bubbles might interact, leading to fascinating possibilities.

CHAPTER 6: THE 'S' ENERGY FIELD EXPLAINED

A New Perspective on Reality

Imagine the universe as an immense ocean, vast and deep. Within this ocean, countless bubbles rise and fall, each one a world unto itself. These bubbles represent independent realms, coexisting within the same space yet remaining distinct due to their unique properties. This is the essence of the Bubble Theory—a new way to understand the fabric of reality and the possible existence of higher dimensions.

In this chapter, we'll explore the Bubble Theory in simple terms, using analogies and examples to make the concept accessible. We'll delve into how these bubbles can exist simultaneously without interfering with one another and what binds them together in the fourth dimension.

The Radio Analogy: Tuning into Different Realities

Consider how a radio works. In any given place, numerous radio signals fill the air, each broadcasting on its own frequency. When you tune your radio to a specific frequency, you access that particular station, while the others remain undetected. All these signals coexist in the same space without interfering because they operate at different frequencies.

Similarly, the Bubble Theory suggests that multiple universes —or "bubbles"—exist within the same spatial framework but are differentiated by unique "frequencies" known as 'S' energy signatures. Just as radio frequencies allow multiple stations to broadcast simultaneously, the 'S' energy signatures enable multiple bubbles to occupy the same space without interacting.

Understanding the 'S' Energy Signature

At the heart of the Bubble Theory is the concept of the 'S' Energy Field and its unique signatures. Think of the 'S' energy signature as a specific vibration or pattern that defines each bubble, much like a fingerprint or a musical note.

- **Unique Identifier:** The 'S' energy signature is what makes each bubble distinct, preventing it from merging with others.
- **Binding Force:** It acts as a glue, holding the matter within a bubble together in the fourth dimension.
- **Non-Interference:** Because each bubble has a unique signature, they don't interfere with one another, allowing for peaceful coexistence.

Bubbles Coexisting in the Fourth Dimension

To visualize how bubbles can occupy the same space without interacting, let's use the analogy of transparent soap bubbles. Imagine blowing several soap bubbles that float and sometimes overlap. Because they're transparent, you can see through them, and when they overlap, each bubble maintains its own surface tension, keeping it intact.

In the Bubble Theory, each bubble represents a universe with its own set of physical laws and matter, bound by its 'S' energy signature. Even if these bubbles overlap in space, their unique signatures prevent them from interacting directly.

Layers of Reality: The Book Pages Analogy

Another way to think about this is to imagine a book. Each page in the book represents a different bubble, a separate reality.

The pages are stacked together, occupying the same overall space (the book), but the characters on one page are unaware of those on another. The 'S' energy signatures are like the page numbers, keeping each story distinct and organized.

Why Don't We Perceive Other Bubbles?

You might wonder why we don't notice these other bubbles if they occupy the same space as ours. The answer lies in our sensory limitations and the uniqueness of the 'S' energy signatures.

- **Sensory Limitations:** Our senses and instruments are tuned to detect phenomena within our own bubble's 'S' energy signature.
- **Different Frequencies:** Just as a radio tuned to one frequency doesn't pick up other stations, we're "tuned" to our bubble's signature.
- **Dimensional Constraints:** Our perception is confined to three spatial dimensions, making it challenging to detect or interact with phenomena in higher dimensions.

An Everyday Example: Multiple Conversations

Imagine being at a crowded party where many conversations happen simultaneously. You can focus on the person you're talking to, while other conversations fade into the background noise. Each conversation is like a bubble with its own 'S' energy signature. While they're all occurring in the same room, they don't interfere with your ability to communicate with your conversation partner.

The Non-Interference Principle

A key aspect of the Bubble Theory is the non-interference principle. Because each bubble has a unique 'S' energy signature, they remain independent even when occupying the same space.

This principle ensures that:

- **Structural Integrity:** Each bubble maintains its own physical laws and structures without being affected by others.
- **Energy Isolation:** Energy and matter within one bubble don't transfer to another under normal circumstances.
- **Stability:** The uniqueness of signatures prevents chaotic interactions that could destabilize the bubbles.

Possibility of Interaction: A Glimpse into the Extraordinary

While the non-interference principle keeps bubbles separate, the Bubble Theory also introduces the intriguing possibility of interaction under special conditions.

- **Interlinking Bubbles:** If two bubbles' 'S' energy signatures become sufficiently similar, they might begin to interact.
- **Energy Fluctuations:** External influences or internal fluctuations could alter a bubble's signature, leading to potential overlap.
- **Phenomenological Implications:** Such interactions could explain certain unexplained phenomena or anomalies in our universe.

We'll explore these ideas further in subsequent chapters, delving into how interlinking might occur and what it could mean for our understanding of reality.

Embracing the Unknown

The Bubble Theory invites us to expand our minds and consider possibilities beyond our everyday experience. By using analogies and familiar concepts, we can begin to grasp how multiple realities might coexist and how the fourth dimension plays a crucial role in this framework.

As we continue our journey, we'll examine the 'S' Energy Field in more detail, explore the nature of bubbles in the fourth dimension, and discuss the implications of this theory for physics and cosmology.

Key Takeaways from Chapter 6:

- **Bubble Theory Overview:** Proposes that multiple universes (bubbles) exist simultaneously in the same space, differentiated by unique 'S' energy signatures.
- **'S' Energy Signature:** Acts as a unique identifier and binding force for each bubble, preventing interference with others.
- **Analogies Used:**
  - **Radio Frequencies:** Different stations (bubbles) coexist without interference due to unique frequencies ('S' energy signatures).
  - **Soap Bubbles:** Overlapping bubbles maintain integrity due to their unique surface tension.
  - **Book Pages:** Separate stories (bubbles) coexist within the same book (space) without characters interacting.
- **Non-Interference Principle:** Ensures that bubbles remain independent even when occupying the same space.
- **Possibility of Interaction:** Under special conditions, bubbles might interact, leading to fascinating possibilities.

CHAPTER 7: BUBBLES IN THE FOURTH DIMENSION

Unveiling the 'S' Energy Field

At the core of the Bubble Theory lies the enigmatic 'S' Energy Field. This field permeates the fourth spatial dimension, acting as the medium through which 'S' energy signatures propagate. Understanding this field is essential to grasping how bubbles form, maintain their integrity, and potentially interact.

What Is the 'S' Energy Field?

- **A Higher-Dimensional Medium:** The 'S' Energy Field exists in the fourth spatial dimension, beyond our direct perception.
- **Carrier of Signatures:** It facilitates the propagation of unique 'S' energy signatures that define each bubble.
- **Analogy to Electromagnetic Field:** Just as the electromagnetic field allows for the transmission of light and radio waves, the 'S' Energy Field enables the existence of bubbles in higher dimensions.

Visualizing the 'S' Energy Field

The Ocean Analogy

Imagine the fourth dimension as a vast ocean—the 'S' Energy Field is the water itself. Within this ocean, waves of various frequencies travel, representing different 'S' energy signatures. Each bubble is like a boat on the ocean, buoyed and moved by its unique wave.

- **Waves as Signatures:** The unique patterns of the

waves correspond to the unique 'S' energy signatures of each bubble.
- **Boats as Bubbles:** Each boat navigates the ocean independently, carried by its wave, much like bubbles exist within the 'S' Energy Field.

Musical Harmony

Consider an orchestra where each instrument plays a different note. The air (analogous to the 'S' Energy Field) carries these notes simultaneously without the sounds interfering destructively. Each note maintains its distinct identity while contributing to the overall harmony.

- **Air as the Medium:** Just as air carries sound waves, the 'S' Energy Field carries 'S' energy signatures.
- **Unique Notes:** Each instrument's note represents a bubble's unique signature.
- **Harmonious Coexistence:** The notes coexist and create harmony, similar to how bubbles coexist within the field.

Properties of the 'S' Energy Field

- **Pervasive:** It exists everywhere in the fourth dimension, underlying all bubbles.
- **Dynamic:** The field can exhibit fluctuations and variations, potentially influencing bubble behavior.
- **Binding Mechanism:** It holds the matter within a bubble together, ensuring structural integrity in higher dimensions.

How 'S' Energy Signatures Bind Matter

The 'S' energy signature of a bubble determines how its matter is organized and maintained.

- **Unique Vibrations:** The signature causes matter within the bubble to vibrate at specific frequencies, aligning it within the 'S' Energy Field.
- **Cohesion:** This alignment creates a cohesive force that binds

particles together, similar to how magnetic fields can align and hold magnetic materials.
- **Stability:** The consistent vibration ensures the bubble's stability within the fourth dimension.

Interplay Between Bubbles and the 'S' Energy Field

Non-Interference Through Orthogonality

In mathematical terms, orthogonality refers to functions or vectors being perpendicular, resulting in zero interaction. In the Bubble Theory:

- **Orthogonal Signatures:** Each bubble's 'S' energy signature is orthogonal to others, preventing interference.
- **Mathematical Parallel:** Just as orthogonal waves don't interfere, bubbles with unique signatures coexist without affecting one another.

Potential for Interaction

While orthogonality maintains separation, certain conditions might lead to interaction:

- **Signature Overlap:** If two bubbles' signatures become less orthogonal (more similar), they might begin to interact.
- **Field Fluctuations:** Changes in the 'S' Energy Field could cause temporary overlaps or resonances between bubbles.

Energy and Frequency: The Connection

Drawing from physics, energy and frequency are closely related. Higher frequency corresponds to higher energy.

- **Bubble Energy Levels:** A bubble's stability depends on its 'S' energy level, which correlates with its signature's frequency.
- **Energy Fluctuations:** Variations in energy could alter a bubble's signature, potentially affecting its interaction with

others.

Everyday Analogies to Clarify Concepts

Wi-Fi Networks

In a crowded area, multiple Wi-Fi networks operate simultaneously without significant interference because they use different channels (frequencies).

- **Networks as Bubbles:** Each network represents a bubble with a unique 'S' energy signature.
- **Channels as Signatures:** The specific frequency channel ensures signals remain distinct.
- **Non-Interference:** Devices connect to their intended network without cross-talk.

Color Frequencies

Visible light comprises different colors, each with its own frequency.

- **Colors as Signatures:** Each color represents a different 'S' energy signature.
- **Light Through a Prism:** A prism separates white light into distinct colors, analogous to distinguishing between bubbles.
- **Coexistence of Colors:** All colors exist within the same beam of light without interfering.

Implications of the 'S' Energy Field

Understanding the 'S' Energy Field opens up possibilities:

- **Explaining Fundamental Forces:** The field might be the underlying mechanism connecting or differentiating fundamental forces.
- **Universe's Structure:** It could influence the large-scale structure of the cosmos, affecting how matter is distributed.
- **New Physics:** Exploring this field might reveal new laws or

principles governing higher-dimensional spaces.

A Bridge Between Dimensions

The 'S' Energy Field serves as a bridge between the familiar three-dimensional world and the fourth spatial dimension. By conceptualizing this field, we take a step closer to understanding how higher dimensions might operate and how they could impact our reality.

Key Takeaways from Chapter 7:

- **'S' Energy Field Defined:** A higher-dimensional field in the fourth spatial dimension that carries unique 'S' energy signatures.
- **Binding Mechanism:** The field binds matter within a bubble, maintaining its structure in higher dimensions.
- **Analogies Used:**
    - **Ocean Waves:** The field as an ocean with waves representing signatures.
    - **Musical Harmony:** The field as air carrying distinct musical notes (signatures).
- **Non-Interference Through Orthogonality:** Unique signatures prevent bubbles from interfering, maintaining independence.
- **Potential for Interaction:** Changes in the field or signatures might allow bubbles to interact.
- **Implications:** Understanding the 'S' Energy Field could provide insights into fundamental forces and the structure of the universe.

CHAPTER 8: WHEN BUBBLES INTERACT

Visualizing the Fourth Dimension

Understanding how bubbles exist and interact in the fourth dimension requires us to stretch our imagination. While we can't perceive the fourth spatial dimension directly, we can use analogies and models to conceptualize it.

The Flatlander Revisited

Recall the Flatlanders from earlier chapters—beings living in a two-dimensional world. To them, a three-dimensional object passing through their plane appears as a series of changing shapes. Similarly, a four-dimensional object interacting with our three-dimensional world might manifest in ways we can't fully comprehend.

Bubbles as Four-Dimensional Structures

In the Bubble Theory:

- **Bubbles Extend into the Fourth Dimension:** They aren't confined to our three-dimensional space but have a structure that includes the fourth spatial dimension.
- **Unique 'S' Energy Signatures:** Each bubble's existence and properties are defined by its signature within the 'S' Energy Field.

Overlapping Spaces Without Interaction

Stacked Sheets Analogy

Imagine a stack of transparent sheets of paper, each representing a different bubble. Though they occupy the same horizontal and vertical space, they remain separate because they're at different "heights"—analogous to positions in the fourth dimension.

- **Independence:** Each sheet has its own drawings (matter and energy) that don't interfere with those on other sheets.
- **Fourth Dimension as Separation:** The "height" separating the sheets represents their unique positions in the fourth dimension, maintained by their 'S' energy signatures.

3D Shadows of 4D Objects

Just as a three-dimensional object casts a two-dimensional shadow, a four-dimensional bubble might cast a three-dimensional "shadow" in our world.

- **Partial Perception:** We might observe effects or phenomena resulting from higher-dimensional structures without seeing them entirely.
- **Unexplained Phenomena:** Certain anomalies or unexplained events could be glimpses of higher-dimensional interactions.

Maintaining Bubble Integrity

Role of the 'S' Energy Signature

The 'S' energy signature ensures that each bubble remains cohesive and distinct.

- **Structural Stability:** It binds matter within the bubble, preventing it from dispersing or merging with other bubbles.
- **Consistency of Physical Laws:** The signature determines the bubble's physical properties and laws, which could differ from those in other bubbles.

Energy Levels and Stability

A bubble's energy level affects its stability in the fourth dimension.

- **Sufficient Energy:** Adequate 'S' energy maintains the bubble's integrity.
- **Energy Fluctuations:** Loss or gain of energy could affect the bubble's stability or cause changes in its 'S' energy signature.

Interaction Between Bubbles

While bubbles generally remain separate due to their unique signatures, certain conditions might allow for interaction.

Interlinking Bubbles

- **Signature Alignment:** If two bubbles' 'S' energy signatures become similar, they may begin to interact.
- **Overlap in the Fourth Dimension:** This could lead to shared spaces or the transfer of energy and matter between bubbles.

Consequences of Interaction

- **Anomalous Events:** Interactions might explain unexplained phenomena, such as sudden appearances or disappearances of matter.
- **Energy Exchange:** Transfer of 'S' energy could impact the stability of the involved bubbles.

Real-World Parallels

Quantum Tunneling

In quantum mechanics, particles can "tunnel" through barriers they shouldn't be able to cross based on classical physics.

- **Analogy to Bubble Interaction:** Just as particles can appear on the other side of a barrier, matter from another bubble might manifest in our bubble under specific conditions.

Resonance in Physics

When two systems share a resonant frequency, they can exchange energy efficiently.

- **Bubbles Aligning:** If bubbles reach a resonant state due to matching 'S' energy signatures, interaction becomes possible.

Challenges in Perception and Detection

Sensory Limitations

Our senses are not equipped to perceive the fourth spatial dimension directly.

- **Instrumental Detection:** We rely on instruments to detect phenomena beyond our natural capabilities, but detecting higher-dimensional effects remains a significant challenge.

Technological Limitations

Current technology may not be sensitive enough to observe or measure interactions with higher-dimensional bubbles.

- **Future Advances:** As technology evolves, we might develop new methods to explore these dimensions.

Embracing Possibility

Exploring bubbles in the fourth dimension encourages us to:

- **Expand Our Horizons:** Consider realities beyond our immediate perception.
- **Foster Innovation:** Inspire new approaches in physics and technology.
- **Cultivate Curiosity:** Keep an open mind about the mysteries of the universe.

Key Takeaways from Chapter 8:

- **Bubbles in 4D:** Bubbles exist as four-dimensional structures

with unique 'S' energy signatures.
- **Coexistence Without Interaction:** Bubbles occupy the same space but remain separate due to their unique signatures and positions in the fourth dimension.
- **Analogies Used:**
  - **Stacked Sheets:** Representing bubbles as layers that don't interact.
  - **Shadows of 4D Objects:** Higher-dimensional objects casting perceivable effects in lower dimensions.
- **Interlinking Possibilities:** Under certain conditions, bubbles might interact, leading to potential energy or matter exchange.
- **Challenges in Detection:** Sensory and technological limitations hinder our ability to perceive or measure higher-dimensional phenomena.
- **Speculative Implications:** Considering how bubble interactions might explain certain unexplained phenomena, though these ideas remain theoretical.

CHAPTER 9: RETHINKING GRAVITY AND THE FORCES

The Phenomenon of Interlinking

Interlinking occurs when bubbles with similar 'S' energy signatures begin to interact. This interaction can lead to fascinating and potentially profound consequences for the bubbles involved.

Conditions for Interlinking

- **Signature Similarity:** The closer the 'S' energy signatures of two bubbles, the higher the likelihood of interaction.
- **Energy Fluctuations:** External forces or internal changes can alter a bubble's signature, bringing it closer to another's.

Mechanisms Leading to Interaction

- **Resonance:** When bubbles reach a resonant frequency, they can exchange energy or matter.
- **Field Disturbances:** Fluctuations in the 'S' Energy Field might cause temporary overlaps between bubbles.

Possible Outcomes of Interaction

Energy Transfer

- **Stabilization or Destabilization:** Energy can flow from one bubble to another, potentially stabilizing one while destabilizing the other.
- **Observable Effects:** Energy transfers might manifest as sudden energy spikes or anomalies within a bubble.

Matter Exchange

- **Transference of Particles:** Particles or objects might move between bubbles, appearing or disappearing inexplicably.
- **Alteration of Physical Laws:** The introduction of matter from another bubble could cause local variations in physical properties.

Merging of Bubbles

- **Partial Merging:** Some regions of bubbles might merge, creating areas where the properties of both bubbles coexist.
- **Complete Merging:** In rare cases, bubbles might fully merge, resulting in a new bubble with combined properties.

Analogies to Clarify Interaction

Tuning Forks and Resonance

Imagine two tuning forks of the same frequency placed near each other. Striking one causes the other to vibrate sympathetically.

- **Bubbles as Tuning Forks:** Similar 'S' energy signatures resonate, leading to interaction.
- **Energy Transfer:** Vibration (energy) moves from one tuning fork (bubble) to the other.

Overlapping Waves

Dropping two stones close together in a pond creates overlapping ripples.

- **Interference Patterns:** Where ripples meet, they interfere constructively or destructively.
- **Bubble Interaction:** Overlapping 'S' energy signatures create areas of interaction.

Potential Real-World Implications

Unexplained Phenomena

- **Disappearances:** Sudden vanishings of objects or even people might be explained by matter transferring to another bubble.
- **Anomalous Events:** Reports of objects appearing from nowhere or changes in physical properties could result from bubble interactions.

Note: These explanations are speculative and not scientifically verified.

Quantum Entanglement Connection

- **Instantaneous Correlations:** Entangled particles might share an 'S' energy signature across bubbles, accounting for their connected behavior.
- **Non-Local Effects:** Interactions in the fourth dimension could provide a framework for understanding quantum non-locality.

Challenges and Considerations

Energy Requirements

- **High Energy Thresholds:** Significant energy might be needed to alter a bubble's 'S' energy signature sufficiently for interaction.
- **Stability Risks:** Manipulating the 'S' Energy Field could risk destabilizing bubbles.

Technological Limitations

- **Current Capabilities:** We lack the technology to intentionally alter 'S' energy signatures or detect interactions reliably.
- **Future Possibilities:** Advancements in science might enable control over bubble interactions.

Ethical and Philosophical Questions

- **Should We Interact?** If interaction becomes possible, ethical considerations arise about the consequences.
- **Impact on Reality:** Altering our bubble or another could have unforeseen effects on both.

Speculative Technologies

Interdimensional Communication

- **Sending Signals:** Developing methods to transmit information between bubbles.
- **Receiving Responses:** Interpreting signals from other bubbles.

Travel Between Bubbles

- **Portal Concepts:** Theoretical gateways allowing matter to move between bubbles.
- **Safety Concerns:** Ensuring stability and safety during transit.

Inspirations from Science Fiction

Many of these ideas have been explored in science fiction, sparking imagination about parallel universes and interdimensional travel.

- **Literature and Film:** Stories often depict characters crossing into alternate realities, reflecting humanity's fascination with the concept.
- **Influence on Science:** Fiction can inspire scientific inquiry, leading to real-world advancements.

Moving Forward with Caution

While the idea of interacting with other bubbles is intriguing, it's essential to approach the concept responsibly.

- **Scientific Rigor:** Any exploration must be grounded in empirical evidence and adhere to scientific methods.

- **Ethical Guidelines:** Consideration of potential impacts is crucial to prevent harm.

Key Takeaways from Chapter 9:

- **Interlinking Defined:** Interaction between bubbles occurs when their 'S' energy signatures become similar.
- **Mechanisms of Interaction:** Resonance and energy fluctuations can lead to bubbles overlapping or exchanging energy and matter.
- **Analogies Used:**
  - **Tuning Forks:** Resonant frequencies leading to energy transfer.
  - **Overlapping Waves:** Interference patterns representing bubble interactions.
- **Potential Implications:** Explaining unexplained phenomena, exploring quantum entanglement, and considering future technologies.
- **Challenges and Ethics:** Energy requirements, technological limitations, and ethical considerations must be addressed.
- **Inspiration from Fiction:** Science fiction explores these themes, reflecting and inspiring real scientific inquiry.

CHAPTER 10: SHEDDING LIGHT ON DARK MATTER AND ENERGY

Introduction

Imagine standing on a trampoline with a massive bowling ball placed at its center. The fabric of the trampoline warps under the weight, creating a dip that influences how smaller balls roll around it. This simple analogy captures the essence of gravity in our three-dimensional world. But what if gravity isn't just a force emanating from mass but a ripple in a higher-dimensional fabric? The Bubble Theory invites us to rethink gravity and the fundamental forces of nature by introducing interactions within the fourth spatial dimension. This chapter explores how gravity could emerge from these higher-dimensional interactions and how this perspective might unify the fundamental forces.

Gravity as an Emergent Phenomenon

Traditional Understanding of Gravity

In classical physics, gravity is described by Isaac Newton as a force of attraction between masses. Later, Albert Einstein revolutionized our understanding with his General Theory of Relativity, portraying gravity not as a force but as a curvature of spacetime caused by mass and energy. Despite these advancements, gravity remains elusive when it comes to unifying with the other fundamental forces: electromagnetism, the weak nuclear force, and the strong nuclear force.

The Bubble Theory's Perspective

The Bubble Theory proposes that gravity is not a fundamental force but an emergent phenomenon resulting from interactions between multiple bubbles in the fourth spatial dimension. Just as ripples on a pond emerge from the movement of objects within the water, gravitational effects in our universe arise from the dynamics of bubbles interacting in higher-dimensional space. This perspective suggests that what we perceive as gravity is a manifestation of these interdimensional interactions.

Unifying Fundamental Forces Through the 'S' Energy Field

One of the most tantalizing prospects of the Bubble Theory is its potential to unify all fundamental forces. The 'S' Energy Field, which governs the interactions of bubbles, serves as a common thread linking gravity with electromagnetism, the weak force, and the strong force. By existing within the fourth dimension, bubbles can influence each other in ways that transcend our three-dimensional understanding, allowing for a cohesive framework where all forces emerge from higher-dimensional dynamics.

Bridging Quantum Mechanics and General Relativity

The Challenge of Unification

Quantum Mechanics excels in explaining the behavior of particles at the smallest scales, while General Relativity provides a robust description of gravity and the large-scale structure of the universe. However, integrating these two pillars of modern physics has proven to be a formidable challenge. The discrepancies between the probabilistic nature of quantum mechanics and the deterministic framework of General Relativity have left physicists searching for a unified theory.

Bubble Theory's Approach

The Bubble Theory offers a novel approach to this unification by positing that both quantum mechanics and gravity emerge from the same higher-dimensional interactions governed by the 'S' Energy Field. In this framework, quantum phenomena like superposition and entanglement are manifestations of bubble interactions in the fourth dimension, seamlessly bridging the gap between the quantum and the gravitational.

Visualizing the Bridge

Consider the analogy of a hologram, where a three-dimensional image is projected from a two-dimensional surface. Similarly, our three-dimensional universe could be a projection of interactions occurring in the fourth dimension. This higher-dimensional activity could give rise to both the familiar forces and the mysterious quantum behaviors, providing a unified tapestry of reality.

Potential Evidence and Observational Signatures

Gravitational Wave Patterns

The Bubble Theory predicts that gravitational waves, ripples in spacetime detected by observatories like LIGO, could carry signatures of higher-dimensional interactions. These waves might exhibit unique patterns or anomalies that deviate from predictions made solely by General Relativity, hinting at the influence of interdimensional dynamics.

Black Hole Behavior

Black holes, with their intense gravitational pull, serve as natural laboratories for testing theories of gravity. The Bubble Theory suggests that the merger of black holes could involve interactions between bubbles in the fourth dimension, potentially leading to observable deviations in the gravitational waves they emit compared to Standard Model predictions.

Cosmic Scale Structures

Large-scale cosmic structures, such as galaxy clusters and voids, could be influenced by the collective behavior of multiple bubbles. Analyzing the distribution and movement of these structures might reveal patterns consistent with higher-dimensional interactions, providing indirect evidence for the Bubble Theory.

Mathematical Framework (Simplified)

Higher-Dimensional Field Equations

At the heart of the Bubble Theory lies a set of extended field equations that incorporate the fourth spatial dimension. These equations describe how the 'S' Energy Field interacts across multiple bubbles, influencing the fundamental forces. While the complete mathematical formulation is intricate, the simplified version captures the essence of these higher-dimensional interactions.

Energy Conservation Across Dimensions

Energy conservation remains a fundamental principle within the Bubble Theory. However, in the context of higher-dimensional interactions, energy exchanges between bubbles adhere to adapted conservation laws that account for the fourth dimension. This ensures consistency with established physical laws while allowing for the emergence of new phenomena.

Quantum Mechanics Integration

Integrating quantum mechanics with the Bubble Theory involves reconciling the probabilistic nature of particle interactions with the deterministic framework of higher-dimensional dynamics. This is achieved through advanced mathematical techniques that bridge the gap between quantum states and bubble interactions, fostering a unified description of reality.

Challenges and Future Research

Empirical Validation

One of the primary challenges for the Bubble Theory is empirical validation. Detecting indirect evidence of higher-dimensional interactions requires sophisticated observational techniques and precise data analysis. Future research must focus on refining these methods to identify the subtle signatures predicted by the theory.

Theoretical Refinement

The mathematical models underpinning the Bubble Theory need further development to provide detailed predictions and align with existing physical laws. Integrating quantum mechanics with higher-dimensional field theories remains a significant hurdle that requires collaborative efforts across multiple disciplines.

Technological Advancements

Advancements in telescope technology, particle detectors, and computational modeling are essential for testing the Bubble Theory's predictions. Investing in next-generation instruments will enhance our ability to observe gravitational wave anomalies, black hole behaviors, and cosmic structures with greater precision.

Interdisciplinary Collaboration

Addressing the complexities of the Bubble Theory necessitates collaboration between physicists, mathematicians, computer scientists, and astronomers. Interdisciplinary research initiatives can foster innovative approaches to solving the theory's challenges and accelerate progress toward validation.

Conclusion

The Bubble Theory offers a transformative perspective on

gravity and the fundamental forces, proposing that what we perceive as gravity emerges from higher-dimensional interactions governed by the 'S' Energy Field. By unifying gravity with electromagnetism, the weak force, and the strong force within a cohesive framework, the theory bridges the gap between quantum mechanics and General Relativity. While significant challenges remain in empirical validation and theoretical refinement, the Bubble Theory stands as a promising avenue for advancing our understanding of the cosmos. Continued research, technological innovation, and interdisciplinary collaboration will be pivotal in uncovering the truths that lie beyond the visible veil.

Key Takeaways from Chapter 10:

- **Gravity as an Emergent Phenomenon:** The Bubble Theory suggests that gravity arises from interactions between bubbles in the fourth spatial dimension, rather than being a fundamental force.
- **Unifying Fundamental Forces:** The 'S' Energy Field serves as a common thread linking gravity with electromagnetism, the weak force, and the strong force, offering a potential unified framework.
- **Bridging Quantum Mechanics and General Relativity:** By positing that both quantum phenomena and gravitational forces emerge from higher-dimensional interactions, the Bubble Theory provides a pathway toward unifying these two pillars of modern physics.
- **Potential Evidence:**
    - **Gravitational Wave Patterns:** Unique signatures in gravitational waves could indicate higher-dimensional influences.
    - **Black Hole Behavior:** Deviations in black hole mergers may reveal interdimensional interactions.
    - **Cosmic Scale Structures:** Anomalies in the distribution and movement of cosmic structures might hint at the collective behavior of multiple bubbles.

- **Mathematical Framework:** Higher-dimensional field equations and adapted energy conservation principles form the mathematical backbone of the theory, facilitating the integration of quantum mechanics with gravitational dynamics.
- **Challenges and Future Research:**
  - **Empirical Validation:** Developing sophisticated observational techniques to detect subtle signatures of higher-dimensional interactions.
  - **Theoretical Refinement:** Enhancing mathematical models to align with existing physical laws and provide detailed predictions.
  - **Technological Advancements:** Investing in next-generation instruments to observe gravitational wave anomalies, black hole behaviors, and cosmic structures with greater precision.
  - **Interdisciplinary Collaboration:** Fostering cooperation between various scientific disciplines to address the multifaceted challenges of the Bubble Theory.

CHAPTER 11: QUANTUM CONNECTIONS

Introduction

Imagine gazing into the night sky, observing countless stars and galaxies, yet feeling a profound sense that something invisible is holding the universe together. This unseen force is what scientists refer to as dark matter, and its mysterious counterpart, dark energy, drives the accelerated expansion of the cosmos. Despite comprising approximately 95% of the universe's total mass-energy content, dark matter and dark energy remain elusive, defying direct detection and challenging our understanding of fundamental physics. Enter the Bubble Theory—a groundbreaking framework that offers innovative explanations for these enigmatic phenomena by positing the existence of multiple, overlapping bubbles within the fourth spatial dimension. This chapter delves into how the Bubble Theory accounts for dark matter and dark energy, reshaping our perception of the universe's composition and expansion.

Overlapping Bubbles as Dark Matter

Mechanism of Overlapping Bubbles

Within the Bubble Theory, our universe is envisioned as one bubble among countless others, each occupying a unique position in the fourth spatial dimension. These bubbles are bound by the 'S' Energy Field, ensuring their stability and distinctiveness. When two or more bubbles overlap within this higher-dimensional space, their 'S' Energy Fields interact, leading to gravitational influences that extend into our observable universe.

This interaction creates what appears to be additional mass—dark matter—affecting the motion of celestial bodies without being directly observable through electromagnetic means.

Gravitational Effects

Traditional explanations for dark matter rely on hypothetical particles that interact primarily through gravity. However, the Bubble Theory offers an alternative perspective: the gravitational anomalies attributed to dark matter are the result of overlapping bubbles exerting additional gravitational pull within our universe. These interactions manifest as unseen mass influencing galaxy rotation curves, gravitational lensing, and the large-scale structure of the cosmos. Unlike ordinary matter, dark matter does not emit, absorb, or reflect light, making it invisible to current detection methods, yet its gravitational effects are indispensable for explaining various cosmic phenomena.

Visualizing the Interaction

Picture two transparent soap bubbles floating in close proximity. When they overlap, their surfaces merge without altering their individual integrity. Similarly, overlapping cosmic bubbles interact gravitationally without merging, maintaining their unique 'S' Energy signatures. This visual analogy helps in understanding how multiple bubbles can coexist and influence each other, giving rise to the gravitational effects we attribute to dark matter.

Impact on the Universe's Expansion

Dark Energy and Cosmic Acceleration

In addition to dark matter, observations have revealed that the universe's expansion is accelerating—a phenomenon attributed to dark energy. Dark energy is thought to permeate all of space, acting as a repulsive force counteracting gravity on cosmic scales. The Bubble Theory provides a compelling explanation for dark energy by linking it to the dynamics of overlapping bubbles within the fourth spatial dimension. As bubbles interact, the cumulative effect of these interactions could drive the accelerated expansion, effectively serving as the repulsive force we observe.

Interplay Between Dark Matter and Energy

The relationship between dark matter and dark energy is complex and not fully understood. The Bubble Theory posits that both dark matter and dark energy arise from the interactions of multiple bubbles within the higher-dimensional space. Dark matter's gravitational pull and dark energy's repulsive force could be two manifestations of the same underlying phenomenon —the collective behavior of overlapping bubbles. This unified perspective not only simplifies the explanations for these mysterious components but also offers a cohesive framework that connects the two.

Visualizing Cosmic Expansion

Imagine a loaf of raisin bread rising in the oven. As the dough expands, the raisins (representing galaxies) move away from each other. In the Bubble Theory, the expansion of our universe is akin to the rising dough, driven by the interactions of bubbles that generate the repulsive force of dark energy. This visualization underscores how interdimensional dynamics can influence the large-scale structure and behavior of the cosmos.

Implications for Cosmology

Structure Formation

Understanding dark matter as a product of overlapping bubbles has profound implications for the formation and distribution of cosmic structures. Dark matter acts as a gravitational scaffold around which visible matter coalesces, forming galaxies and clusters. If overlapping bubbles provide the necessary gravitational influence, this offers a natural explanation for the observed distribution and density of dark matter across the universe. It also aligns with simulations that require dark matter to account for the large-scale structure of the cosmos.

Galaxy Rotation Curves

One of the pivotal pieces of evidence for dark matter comes from the rotation curves of galaxies. Observations show that stars at the edges of galaxies orbit at similar speeds to those near the center, defying Newtonian predictions based solely on visible matter. The Bubble Theory explains this anomaly by attributing the additional gravitational pull to overlapping bubbles. This eliminates the need for unseen particles, offering a geometric and interdimensional explanation rooted in higher-dimensional interactions.

Cosmic Microwave Background (CMB) Distortions

The Cosmic Microwave Background provides a snapshot of the early universe, containing subtle fluctuations that inform our understanding of cosmic composition. The Bubble Theory suggests that interactions between bubbles during the universe's formative years could leave imprints on the CMB, resulting in specific distortion patterns. Detecting such anomalies would offer compelling evidence for the theory, linking dark matter and dark energy to higher-dimensional dynamics.

Potential Evidence and Observational Signatures

Gravitational Lensing Anomalies

Gravitational lensing, the bending of light around massive objects, serves as an indirect method for detecting dark matter. The Bubble Theory predicts that overlapping bubbles would create distinct lensing patterns, potentially identifiable through precise astronomical observations. Deviations from expected lensing models could indicate the presence of additional gravitational sources stemming from higher-dimensional interactions.

Cosmic Microwave Background (CMB) Distortions

Interactions between bubbles during the early universe could leave subtle imprints on the CMB. These distortions might manifest as irregularities or unexpected fluctuations in the temperature and polarization of the CMB. Advanced measurements and analyses could reveal patterns consistent with the predictions of the Bubble Theory, providing indirect evidence of higher-dimensional interactions.

High-Energy Particle Fluxes

Particle accelerators and cosmic ray detectors continuously probe the high-energy environment of the universe. The Bubble Theory suggests that energy exchanges between overlapping bubbles could result in unusual energy distributions or exotic particles that defy Standard Model explanations. Identifying and analyzing these high-energy events could serve as tangible signs of higher-dimensional interactions.

Mathematical Framework (Simplified)

Higher-Dimensional Geometry

The interactions of overlapping bubbles are modeled using higher-dimensional geometry, extending beyond the familiar three spatial dimensions. Concepts such as hyperspheres and hypercubes are employed to describe the shape and behavior of bubbles within the fourth dimension. This geometric approach allows for the calculation of gravitational influences resulting from bubble overlaps, providing a mathematical foundation for the theory's explanations.

'S' Energy Field Equations

The 'S' Energy Field is governed by extended field equations that incorporate higher-dimensional parameters. These equations describe how the field propagates through the fourth dimension and interacts with multiple bubbles. Solving these equations under specific conditions predicts the gravitational and energetic effects observed in our universe, offering a quantifiable basis for the theory's claims.

Energy Conservation and Distribution

Energy conservation principles are adapted to higher-dimensional contexts within the Bubble Theory. Energy exchanges between bubbles adhere to these adapted laws, ensuring that the total energy remains consistent across dimensions. The distribution of energy resulting from overlapping interactions can be quantified, providing a basis for comparing theoretical predictions with observational data.

Challenges and Future Research

Empirical Validation

One of the foremost challenges for the Bubble Theory is empirical validation. Detecting indirect evidence of overlapping bubbles requires advanced observational techniques and precise data analysis. Future research must focus on refining these methods to identify the subtle signatures predicted by the theory, such as specific patterns in gravitational lensing or CMB distortions.

Theoretical Refinement

The mathematical models underpinning the Bubble Theory need further development to provide detailed predictions and align with existing physical laws. Integrating quantum mechanics with higher-dimensional field theories remains a significant hurdle that requires collaborative efforts across multiple disciplines. Enhancing the precision and predictive power of these models will facilitate their integration with established frameworks in physics.

Technological Advancements

Advancements in telescope technology, particle detectors, and computational modeling are essential for testing the Bubble Theory's predictions. Investing in next-generation instruments will enhance our ability to observe gravitational lensing anomalies, CMB distortions, and high-energy particle fluxes with greater precision, thereby providing the necessary tools for empirical validation.

Interdisciplinary Collaboration

Addressing the complexities of the Bubble Theory necessitates collaboration between physicists, mathematicians, astronomers, and computer scientists. Interdisciplinary research initiatives can foster innovative approaches to solving the theory's multifaceted challenges and accelerate progress toward validation.

Conclusion

The Bubble Theory offers a groundbreaking perspective on dark matter and dark energy, proposing that these elusive components of the universe arise from overlapping bubbles within the fourth spatial dimension. By providing geometric and dynamic explanations rooted in higher-dimensional interactions, the theory unifies previously disparate phenomena under a single framework. While significant challenges remain in empirical validation and theoretical refinement, the Bubble Theory stands as a promising avenue for advancing our understanding of the cosmos. Continued research, technological innovation, and interdisciplinary collaboration will be pivotal in uncovering the truths that lie beyond the visible veil.

CHAPTER 12: PHILOSOPHICAL REFLECTIONS

Introduction

Imagine you're in a bustling café, observing two friends engaged in a deep conversation. Despite being in the same room, their interaction seems effortless, almost instantaneous, as if they share a secret language. This scenario mirrors the intriguing phenomenon of quantum entanglement, where particles become interconnected in ways that defy our classical understanding of space and time. But what if the key to unraveling this mystery lies beyond our three-dimensional perception? Enter the Bubble Theory—a groundbreaking framework that bridges quantum mechanics and higher-dimensional interactions, offering a fresh perspective on the enigmatic connections that govern the quantum realm.

Understanding Quantum Entanglement

The Basics of Quantum Mechanics

At the heart of quantum mechanics lies the principle that particles can exist in multiple states simultaneously—a concept known as superposition. Unlike classical objects that occupy definite positions and velocities, quantum particles embody probabilities until measured. This probabilistic nature is encapsulated in the wavefunction, a mathematical description of a particle's quantum state.

What Is Quantum Entanglement?

Quantum entanglement occurs when two or more particles

become so deeply linked that the state of one instantly influences the state of the other, regardless of the distance separating them. This phenomenon challenges our intuitive understanding of locality, where objects are only directly influenced by their immediate surroundings. Einstein famously referred to entanglement as "spooky action at a distance," highlighting its counterintuitive nature.

Non-Locality and Instantaneous Connections

Entangled particles exhibit correlations that persist even when separated by vast distances. When a measurement is performed on one particle, the corresponding property of its entangled partner is instantly determined, defying the speed of light constraint imposed by relativity. This non-local behavior suggests that entangled particles share a deeper connection that transcends our conventional notions of space and time.

The Bubble Theory's Perspective on Quantum Entanglement

Higher-Dimensional Interactions

The Bubble Theory posits that our universe exists as a bubble within a higher-dimensional space governed by the 'S' Energy Field. Within this framework, quantum entanglement is reimagined as an interaction facilitated by these higher dimensions. When particles become entangled, it's not just a phenomenon occurring within our three-dimensional space but also involving connections across the fourth spatial dimension.

Non-Locality Explained by Higher Dimensions

In the context of the Bubble Theory, the instantaneous connections observed in entangled particles are a result of their shared 'S' energy signatures in the fourth dimension. Think of it as two friends communicating through a hidden tunnel that bypasses the need to traverse the three-dimensional space between them. This higher-dimensional bridge allows for instantaneous information transfer without violating the speed of light constraint in our observable universe.

Unified Framework for Quantum and Gravitational Phenomena

By integrating quantum mechanics with higher-dimensional interactions, the Bubble Theory provides a unified framework that reconciles the probabilistic nature of quantum entanglement with the deterministic behavior of gravitational forces. This unification is a significant stride toward a comprehensive understanding of the fundamental forces that shape our universe.

Implications of Quantum Connections in the Bubble Theory

Revisiting the EPR Paradox

The Einstein-Podolsky-Rosen (EPR) paradox challenges the completeness of quantum mechanics by highlighting the strange implications of entanglement. The Bubble Theory offers a resolution by positing that the seemingly instantaneous connections are a result of higher-dimensional interactions, maintaining the integrity of both quantum mechanics and relativity.

Potential for Faster-Than-Light Communication

While quantum entanglement appears to allow for instantaneous connections, it doesn't permit faster-than-light communication in the traditional sense because the outcome of measurements is inherently random. However, understanding entanglement through the Bubble Theory could inspire new ways of thinking about information transfer and connectivity across dimensions, potentially unlocking novel communication paradigms.

Influence on Quantum Computing and Information

Quantum computing leverages entanglement to perform complex computations more efficiently than classical computers. The Bubble Theory's insights into higher-dimensional interactions could lead to advancements in quantum information processing, enhancing the capabilities and stability of quantum computers by harnessing higher-dimensional resources.

Mathematical Framework (Simplified)

Entanglement Entropy in Higher Dimensions

Entanglement entropy measures the degree of entanglement between subsystems. In the Bubble Theory, this concept extends into higher dimensions, allowing for a more nuanced calculation that accounts for interactions beyond the three familiar spatial axes. This extension provides a mathematical foundation for understanding how higher-dimensional dynamics influence quantum states.

'S' Energy Field and Quantum States

The 'S' Energy Field governs the interactions between bubbles in higher-dimensional space. Quantum states within a bubble are influenced by the field's dynamics, providing a mechanism for entanglement that transcends traditional three-dimensional constraints. This relationship is encapsulated in extended field equations that describe how the 'S' Energy Field propagates and interacts with quantum particles.

Probability Amplitudes and Dimensional Influence

Probability amplitudes, fundamental to quantum mechanics, describe the likelihood of a particle's state. The Bubble Theory modifies these amplitudes by incorporating higher-dimensional factors, offering a more detailed prediction of entangled states and their behaviors. This modification enhances our understanding of quantum phenomena by embedding them within a higher-dimensional framework.

Challenges and Future Research

Empirical Validation of Quantum Connections

While the Bubble Theory offers a compelling framework, empirical validation remains a significant challenge. Detecting indirect evidence of higher-dimensional interactions influencing quantum entanglement requires sophisticated experimental setups and precise data analysis. Future research must focus on designing experiments that can identify subtle signatures predicted by the theory, such as specific patterns in entanglement entropy or anomalies in quantum state behaviors.

Integrating with Existing Quantum Theories

Ensuring compatibility between the Bubble Theory and established quantum theories requires meticulous mathematical refinement and theoretical exploration. This integration is essential for the theory's acceptance within the scientific community and for bridging gaps between disparate areas of physics.

Technological Innovations for Higher-Dimensional Probing

Advancements in technology, such as higher-dimensional particle detectors and quantum entanglement measurement devices, are necessary to explore the implications of the Bubble Theory. Investing in these technologies will facilitate deeper insights into quantum connections and validate the theory's predictions.

Interdisciplinary Collaboration

Bridging the gap between quantum physics and higher-dimensional theories necessitates collaboration across multiple disciplines. Physicists, mathematicians, and computer scientists must work together to develop comprehensive models and experimental frameworks that can test the Bubble Theory's predictions.

Conclusion

Quantum entanglement remains one of the most fascinating and mysterious aspects of quantum mechanics. The Bubble Theory provides a novel perspective by introducing higher-dimensional interactions as the underlying mechanism for entanglement and non-locality. By reconciling the probabilistic nature of quantum mechanics with the deterministic framework of higher-dimensional physics, the Bubble Theory offers a unified approach that could pave the way for groundbreaking advancements in our understanding of the universe. While significant challenges remain in validating and integrating this theory, the potential rewards in terms of scientific discovery and technological innovation are immense.

CHAPTER 13: THE JOURNEY OF DISCOVERY

Redefining Reality

Have you ever pondered the nature of reality and our place within it? The Bubble Theory challenges conventional notions, inviting us to envision a universe that extends beyond the three-dimensional fabric we perceive. This chapter delves into the philosophical implications of the Bubble Theory, exploring how it redefines our understanding of existence, consciousness, and the very essence of reality.

The Nature of Reality in Higher Dimensions

Beyond the Observable Universe

Our perception of reality is confined to the three spatial dimensions, yet theories like the Bubble Theory suggest the existence of higher-dimensional spaces. These additional dimensions offer a canvas where multiple universes coexist, interact, and influence each other in ways that transcend our everyday experiences.

Perception and Consciousness

If higher dimensions influence our universe, what does that mean for our consciousness and perception? The Bubble Theory posits that our awareness operates within the three-dimensional constraints, yet it could be subtly influenced by the dynamics of the fourth dimension. This interplay raises questions about the potential for consciousness to tap into higher-dimensional information or experiences.

Existence and Free Will

In a universe where multiple bubbles interact and overlap, the concepts of destiny and free will take on new dimensions. Are our choices influenced by higher-dimensional interactions? Does the overlap of bubbles allow for alternate realities where different choices are made? These philosophical inquiries challenge us to rethink the foundations of autonomy and determinism.

Consciousness and the 'S' Energy Field

The Interface Between Mind and Matter

The 'S' Energy Field, as the governing force of higher-dimensional interactions, could play a role in bridging the gap between consciousness and the physical world. Some philosophical interpretations suggest that consciousness might interact with the 'S' Energy Field, influencing or being influenced by higher-dimensional dynamics.

Potential for Higher-Dimensional Awareness

If consciousness interacts with higher dimensions, this opens the door to experiences and knowledge beyond our current understanding. The Bubble Theory encourages us to explore the possibilities of higher-dimensional awareness, where the mind could perceive or influence aspects of reality that are invisible to our three-dimensional senses.

Implications for Human Experience

Understanding consciousness within the framework of the Bubble Theory could revolutionize our perception of self and reality. It invites us to consider that our consciousness might be more interconnected with the universe than previously thought, potentially influencing the very fabric of higher-dimensional interactions.

Implications for Free Will

Determinism vs. Free Will

The existence of higher dimensions and multiple interacting bubbles introduces complexities to the age-old debate of determinism versus free will. If higher-dimensional interactions influence our universe, they could affect the outcomes of events in ways that challenge the notion of free agency.

Multiple Realities and Choices

The Bubble Theory's framework suggests the possibility of alternate realities where different choices lead to different outcomes. This multiplicity of bubbles could imply that every decision spawns a new bubble, each representing a different path taken. This concept blurs the lines between fate and free will, presenting a universe where all possibilities coexist.

Empowering Human Agency

Conversely, the awareness of higher-dimensional influences could empower individuals to exert more control over their lives. Understanding the interplay between bubbles might provide insights into how to navigate or influence higher-dimensional dynamics, enhancing personal agency and decision-making.

Cultural and Societal Impact

Representation in Literature and Art

The Bubble Theory resonates with themes explored in literature and art, where alternate realities, parallel universes, and higher-dimensional spaces are common motifs. From the multiverse in science fiction to abstract art depicting higher dimensions, cultural expressions reflect our innate curiosity about the unknown and the unseen.

Shaping Philosophical Thought

Philosophical discourse has long grappled with the nature of reality, consciousness, and existence. The Bubble Theory contributes to this dialogue by providing a scientific framework that aligns with and challenges existing philosophical paradigms, fostering interdisciplinary discussions that bridge science and philosophy.

Influence on Scientific Paradigms

As theories like the Bubble Theory gain traction, they influence the direction of scientific research and inquiry. By proposing higher-dimensional interactions as a foundational aspect of reality, the theory encourages scientists to explore new methodologies, technologies, and conceptual frameworks that transcend traditional boundaries.

Conclusion

The Bubble Theory invites us to expand our minds beyond the familiar three dimensions, challenging us to envision a universe where higher-dimensional interactions influence the very fabric of reality. By redefining our understanding of consciousness, existence, and free will, the theory bridges the gap between scientific inquiry and philosophical exploration. As we continue to delve deeper into these higher dimensions, we open doors to new realms of knowledge and possibilities, reshaping our perception of ourselves and the universe we inhabit.

CHAPTER 14: THE FUTURE HORIZONS

Introduction

Have you ever embarked on a journey with no clear destination, driven solely by curiosity and the desire to uncover the unknown? The pursuit of scientific understanding often mirrors such a voyage—filled with exploration, challenges, and moments of revelation. In the realm of theoretical physics, developing groundbreaking ideas like the Bubble Theory requires not only intellectual rigor but also a profound sense of wonder and perseverance. This chapter delves into the personal and professional journey behind the Bubble Theory, highlighting the motivations, challenges, and inspirations that have shaped its evolution.

Personal Motivation

Inspiration from the Cosmos

From a young age, gazing up at the night sky instilled a sense of awe and curiosity about the universe's vastness and mysteries. The shimmering stars and the intricate dance of celestial bodies sparked questions: What lies beyond our visible horizon? Are there realities parallel to our own? These contemplations laid the foundation for a lifelong quest to understand the fundamental nature of reality.

The Quest for Unity in Physics

One of the most compelling motivations behind the Bubble Theory is the desire to unify the fundamental forces of nature.

The apparent disconnect between quantum mechanics and general relativity has been a persistent challenge in physics. The aspiration to bridge this gap—a cornerstone of theoretical physics—fueled the exploration of higher-dimensional interactions and the formulation of the 'S' Energy Field concept.

A Desire to Simplify Complexity

The universe's complexity can be overwhelming, but there is beauty in seeking simplicity amidst intricacy. The Bubble Theory aims to distill complex phenomena like dark matter, dark energy, and quantum entanglement into a cohesive framework that is both elegant and accessible. This drive to simplify without sacrificing depth has been a guiding principle throughout the development of the theory.

The Scientific Process

Formulating the Hypothesis

Every scientific journey begins with a hypothesis—a tentative explanation for observed phenomena. For the Bubble Theory, the hypothesis emerged from pondering the limitations of existing models and envisioning a higher-dimensional framework that could account for unresolved mysteries. This initial idea served as the seed from which the theory would grow.

Mathematical Modeling and Simulations

Translating conceptual ideas into mathematical models is a critical step in validating any theory. Rigorous mathematical formulations were developed to describe the interactions of bubbles within the fourth spatial dimension. Computational simulations played a pivotal role in testing these models, allowing for the visualization of higher-dimensional dynamics and the prediction of observable effects.

Collaborative Efforts

Science thrives on collaboration and the exchange of ideas. Engaging with fellow physicists, mathematicians, and interdisciplinary scholars provided invaluable feedback and perspectives that enriched the Bubble Theory.

Overcoming Challenges

Navigating Uncharted Territories

Venturing into higher-dimensional theories presents unique challenges, both conceptually and mathematically. The lack of empirical data and the abstract nature of the concepts necessitated innovative thinking and perseverance. Overcoming these hurdles required a balance of creativity and discipline, pushing the boundaries of conventional scientific inquiry.

Balancing Simplicity and Complexity

Striking the right balance between simplicity and complexity is an ongoing challenge. While the goal is to make the theory accessible, it must also account for the intricate behaviors observed in the universe. This balance was achieved by focusing on foundational principles that could elegantly explain multiple phenomena without becoming overly convoluted.

Encouraging Exploration

Fostering Curiosity in Future Generations

One of the most rewarding aspects of developing the Bubble Theory is the potential to inspire future scientists and thinkers. By presenting complex ideas in an engaging and accessible manner, the theory serves as a catalyst for curiosity and innovation, encouraging others to explore the unknown and contribute to the collective understanding of the universe.

Interdisciplinary Research Opportunities

The Bubble Theory opens avenues for interdisciplinary research, bridging gaps between physics, mathematics, computer science, and philosophy. This cross-pollination of ideas fosters a holistic approach to solving complex problems, leading to breakthroughs that transcend traditional disciplinary boundaries.

Promoting Scientific Literacy

In an age where scientific literacy is crucial, the Bubble Theory aims to make advanced theoretical concepts understandable to a broader audience. By demystifying high-level physics, the theory empowers individuals to appreciate the beauty and complexity of the universe, fostering a more scientifically informed society.

Conclusion

The journey of discovering and formulating the Bubble Theory has been one of exploration, perseverance, and unwavering curiosity. From the initial sparks of inspiration drawn from the cosmos to the meticulous development of mathematical models, each step has contributed to a deeper understanding of the universe's fabric. By overcoming challenges and fostering interdisciplinary collaboration, the Bubble Theory stands as a testament to the power of human inquiry and imagination. As we continue to push the boundaries of what we know, the theory invites us to embrace the unknown, encouraging a continuous quest for knowledge and the ever-expanding horizons of discovery.

GLOSSARY

- **'S' Energy Field:** A higher-dimensional field in the fourth spatial dimension that carries unique energy signatures, binding matter within each bubble and preventing interference between bubbles.
- **Bubble:** An independent energy-matter unit existing within the fourth spatial dimension, characterized by a unique 'S' energy signature.
- **Fourth Dimension:** An additional spatial dimension beyond the three we experience daily, facilitating higher-dimensional interactions and the coexistence of multiple bubbles.
- **Entanglement:** A quantum phenomenon where particles become linked, such that the state of one instantly influences the state of another, regardless of distance.
- **Gravitational Lensing:** The bending of light from distant objects around massive foreground objects, used as evidence for the presence of dark matter.
- **Quantum Tunneling:** A quantum phenomenon where particles pass through energy barriers they classically shouldn't be able to, due to their wave-like properties.
- **Resonance:** The amplification of a system's oscillation when it is subjected to an external force at a matching frequency.
- **Spacetime:** The four-dimensional continuum combining the three spatial dimensions with time, as described in Einstein's theory of relativity.

FURTHER READING AND RESOURCES

Books:

- *The Elegant Universe* by Brian Greene
- *Hyperspace* by Michio Kaku
- *Flatland* by Edwin A. Abbott
- *Quantum Enigma: Physics Encounters Consciousness* by Bruce Rosenblum and Fred Kuttner
- *A Brief History of Time* by Stephen Hawking

Articles and Papers:

- "Quantum Entanglement and Beyond" – *Nature Reviews Physics*
- "Dark Matter and Dark Energy" – *Scientific American*
- "The Quest for Unifying Theories" – *Physics Today*

Online Lectures and Videos:

- TED Talk: "The Hidden Reality" by Brian Greene
- Khan Academy courses on quantum mechanics and general relativity
- YouTube Channels:
    - PBS Space Time
    - Kurzgesagt – In a Nutshell

Educational Websites:

- arXiv.org
- NASA.gov
- Physics.org

JAMEELCHAMBERLAIN

ABOUT THE AUTHOR

Jameel Chamberlain

Jameel Chamberlain is an independent researcher and theorist passionate about exploring the frontiers of physics and cosmology. With a background in theoretical physics, Jameel has dedicated his time to understanding the complex interplay between dimensions, energy fields, and the fundamental forces that govern the universe. Inspired by the works of renowned physicists and driven by a desire to unify disparate aspects of modern physics, Jameel developed the Bubble Theory—a novel framework that introduces higher-dimensional interactions to explain some of the universe's most elusive phenomena. His work bridges the gap between cutting-edge scientific research and accessible, engaging storytelling, making complex theories understandable to both scientific communities and the general public. Beyond his theoretical pursuits, Jameel is an avid reader and communicator, committed to fostering curiosity and inspiring the next generation of scientists. He believes that imagination and creativity are essential components of scientific discovery and strives to make the wonders of the universe accessible to all. When he's not delving into the mysteries of higher dimensions, Jameel enjoys stargazing, reading science fiction, and engaging in philosophical discussions about the nature of reality and existence.

ACKNOWLEDGEMENT

Creating this book has been a journey filled with curiosity, dedication, and the relentless pursuit of understanding. I am profoundly grateful to everyone who has supported and inspired me along the way.

To the Scientific Community

Your relentless pursuit of knowledge, innovative theories, and unwavering dedication to uncovering the universe's secrets have been a constant source of inspiration. The works of pioneers like Albert Einstein, Brian Greene, and Michio Kaku have profoundly influenced my thinking and shaped the development of the Bubble Theory.

To Educators and Mentors

I extend my deepest gratitude to my educators and mentors who nurtured my curiosity and instilled in me a passion for science. Your encouragement and guidance have been the foundation upon which this theory was built. You taught me to ask questions, challenge assumptions, and strive for excellence in every endeavor.

To the Creative Minds

To the artists, writers, and creative thinkers whose works have bridged the gap between complex scientific concepts and accessible storytelling—thank you. Your ability to visualize and communicate abstract ideas has inspired the analogies and narratives that make the Bubble Theory accessible to a broader audience.

To My Family and Friends

Your unwavering support and understanding have been my anchor throughout this journey. Your patience, encouragement, and belief in my work have provided the strength needed to navigate the challenges and triumphs of developing this theory.

To the Readers

Finally, to you, the reader, thank you for embarking on this journey with me. Your curiosity and willingness to explore the unknown are what make this endeavor meaningful. May the Bubble Theory ignite your imagination, deepen your understanding, and inspire you to continue seeking the wonders of the universe.

www.ingramcontent.com/pod-product-compliance
Lightning Source LLC
Chambersburg PA
CBHW050316230526
45471CB00005B/2210